POLAR PROFILES

Terra Nova

Scott's Last Expedition

Beau Riffenburgh

Terra Nova: the British Antarctic Expedition 1910–13

(Scott's Last Expedition)

'The main object of this Expedition,' *The Times* announced to the public of Britain, 'is to reach the South Pole and to secure for the British Empire the honour of this achievement.'
The timing for such a blatantly nationalistic declaration could not have been more appropriate.
It was 12 September 1909, only a few days after a pair of American adversaries, Frederick A. Cook and Robert E. Peary, had each claimed to have attained the North Pole.

▲ Scott with his party at the South Pole, the dissapointment clear on their faces. Left to right – standing: Oates, Scott, Evans. Seated: Bowers, Wilson. Photographer: Henry Bowers. SPRI P48281/12

► The *Terra Nova* and a berg at ice-foot. 16 Jan 1911. Photographer: Herbert Ponting. SPRI P2005/5/196

The international press was brimming with news about the growing row between the two men. And there, in the midst of it all, a Royal Navy Captain – Robert Falcon Scott – was able to steal part of the North Pole thunder by announcing *his* intention of being the first man ever to reach the point on the planet diametrically opposed to that location. He would, he promised, conquer the last place on Earth.

It was not, in fact, a plan that had only recently materialised or that was driven by events in the Arctic. Rather, the concept had played on Scott's hopes and dreams for several years. Scott had first received international acclaim in his role as commander of the British National Antarctic Expedition (1901–04) on the ship *Discovery*. On that venture, his men had conducted the most extensive scientific programme yet run in the Antarctic, and had made several exploratory journeys into regions never before reached. He personally had led a sledging party to the farthest south ever achieved by man – 82°17'S – deep in the heart of that natural wonder of the world, the Great Ice Barrier (now known as the Ross Ice Shelf).

Only two years after *Discovery* returned north, Scott contacted one of his former officers – Michael Barne – about joining him on another southern adventure. He then initiated discussions with the Royal Geographical Society about the idea in January 1907, but before he could state his intentions publicly, another former subordinate – the charming, charismatic Ernest Shackleton – launched his own expedition, with the stated intent of reaching the Pole. For more than a year and a half, Scott waited anxiously to discover if, on his rival's return, the South Pole would still be virgin territory.

Much to Scott's delight, although Shackleton made a marvellous journey, discovered a path to the high Antarctic Plateau in the centre of the continent, and reached a record farthest south of 88°23'S, only 97 geographical miles (180 km) from the Pole, that most southerly goal remained intact. So, while Shackleton was feted and honoured, Scott continued to plan his own assault on the South Pole.

◄ Ernest Shackleton prior to the British Antarctic Expedition. His good looks would combine with his breezy, naturally easy manner to make him a favourite with audiences around the world.

Although it has sometimes since been disputed, those 97 uncovered miles were central to Scott's agenda. As he indicated to Leonard Darwin, the president of the Royal Geographical Society: 'I believe that the main object, that of reaching the South Pole, will appeal to all our countrymen as the one rightly to be pursued at this moment.' Despite a contrary feeling among much of the hierarchy of the British learned societies – which emphasised the scientific research that could be carried out – Scott was clearly interested in the fame and potential fortune that might come from being the first man to the Pole. However, it is also true that he had other aims – he had been very involved in the scientific studies during his first expedition and he was keen to instigate another serious research programme, including investigating the meteorology, magnetism, geology, and zoology of the far south. He also wanted to conduct new geographical exploration, particularly of King Edward VII Land, a region not yet reached because of heavy ice conditions. Therefore, depending upon the particular audience for his fund-raising effort, he highlighted either exploration or science as the expedition's primary goal.

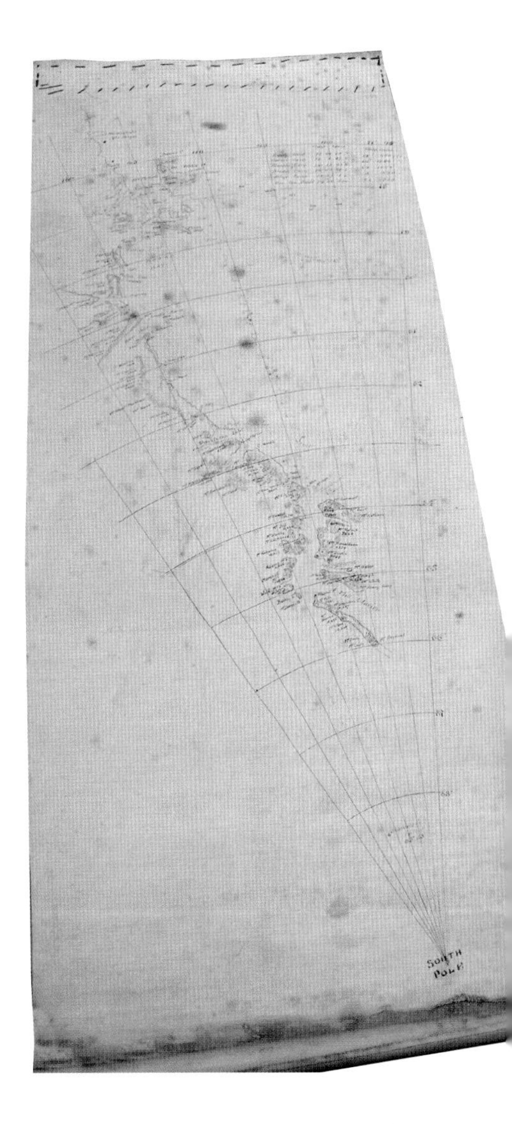

► Copy of Sir Ernest Shackleton's chart of the route to South Pole prepared by Frank Debenham.
SPRI Museum N:1035

Preparations

Although raising funds was an agonisingly slow process, a £20,000 grant from the government enabled the whaler *Terra Nova*, which had been part of the relief effort on Scott's first expedition, to be purchased and refurbished as the expedition ship. At the same time, supplies, equipment and various forms of transport were obtained.

▲ P. O. Edgar Evans. Photograper: Herbert Ponting. SPRI P2005/5/980

Scott had decided that several methods of transport would be used. Much of his hope was placed on tracked motor sledges, which Commander Reginald Skelton, the engineer on *Discovery*, had spent two years designing, developing and testing. Following Shackleton's lead, Scott also planned to take Manchurian ponies, despite the numerous difficulties that Shackleton's expedition had experienced with them in the Antarctic. At the urging of the great Norwegian explorer Fridtjof Nansen – whose unrivalled success in the Arctic had made him the oracle that every serious polar explorer was required to consult – Scott also agreed to take skis. Although his lack of success with dogs on his first expedition meant that he was less than enthusiastic about Nansen's recommendation for their use, he did ultimately agree to include dogs in his planning. Finally, Scott expected – and looked forward to – man-hauling as a major part of the polar journey.

◀ The *Terra Nova* held up in the Pack. 13 December 1910. Photographer: Herbert Ponting. SPRI P2005/5/33

The announcement of the expedition saw more than 8000 applications pour in. Five members of the crew of *Discovery* were brought back: Petty Officers Tom Crean, Edgar Evans, William Heald and Thomas Williamson, and Chief Stoker Bill Lashly. Also returning was Dr Edward Wilson, one of Scott's closest friends, who was named chief of the scientific staff, zoologist, and expedition artist. Meteorologist George Simpson, another would-be member of *Discovery*, who had not been approved for service for that earlier expedition by the Admiralty's medical board, was finally welcomed aboard on secondment from the Indian Weather Bureau.

With the assistance of T.W. Edgeworth David, the renowned professor of geology at the University of Sydney and the effective head of science on Shackleton's expedition, Scott and Wilson assembled an impressive group of scientists. Unfortunately, despite extended efforts, they were not able to entice Douglas Mawson, who

◀ Group of all the shore party (except Clissold who was ill and Mr H. G. Ponting photographing the party) October 1911. Photographer: Herbert Ponting.
SPRI P2005/5/0577

had been another important cog in Shackleton's expedition, to join them – instead Mawson organised his own expedition, which would prove to be an even more successful scientific endeavour.

The geological team was the largest scientific contingent, consisting of three of David's former students or collaborators: Griffith Taylor, Frank Debenham and Raymond Priestley. There were also two biologists: Edward Nelson, who would join the land party, and Dennis Lillie, who would remain aboard *Terra Nova* to carry out oceanographic studies. The physicist was the Canadian Charles Wright, who, in the company of Taylor, had walked from Cambridge to the expedition office in London to convince Scott and Wilson that he was the right man for the job. Finally, the assistant zoologist was Apsley Cherry-

▼ Thermometer and aneroid barometer used on the *Terra Nova* Expedition.
SPRI Museum

▲ Captain L.E.G. Oates. Photographer: Herbert Ponting. SPRI P2005/5/1403

Garrard, the cousin of Reginald Smith (Scott's friend and publisher) and a protégé of Wilson. Cherry-Garrard's contribution of £1000 towards the expedition also helped to convince Scott to take him on.

Financial issues also led to two other men being added to the expedition. One was Captain L.E.G. 'Titus' Oates of the 6th Inniskilling Dragoons, who also paid £1000 to join. His expertise with horses made him a natural choice to be put in charge of the ponies. The other was Lieutenant E.R.G.R. 'Teddy' Evans, who had been the second officer on *Morning* during the *Discovery* relief expeditions. Evans had been organising his own expedition before Sir Clements Markham convinced him to join forces with Scott. In return for being named second-in-command, Evans abandoned his own plans and turned over to Scott all the funds that he had already raised. There would, however, prove to be a major downside to the arrangement. Once in place, Evans told Scott that he did not feel it appropriate that Skelton, who outranked him, should be serving under a junior officer. In response, Scott released Skelton, who had been more or less promised that he would be the second-in-command. He was also the only man who fully understood the motor sledges. His successor as engineer, Bernard Day – who had been in charge of the motor-car that Shackleton had unsuccessfully used on the *Nimrod* expedition – would never be able to coax the production out of the sledges that Skelton felt was possible.

Three other key officers joined the ship some six weeks before she was due to sail: Royal Navy Lieutenant Victor Campbell, the acting first mate; Lieutenant Henry Bowers of the Royal Indian Marine, the officer in charge of supplies; and Royal Navy Lieutenant Harry Pennell, the navigator, who would ultimately take command of *Terra Nova*.

The Voyages

On 15 June 1910, *Terra Nova*, under the command of Evans, sailed from Cardiff, where she had put in to coal and as thanks for the large contributions raised there, while Scott remained in England to raise more money. Two months later, when the expedition reached Simonstown, South Africa, Scott, often called 'the Owner,' assumed command, and began to determine who would join the land party and who would remain with the ship.

▲ *Aug. 8. 1911. 3pm. Cape Evans. Looking north.* Watercolour by Edward Wilson.
SPRI Museum N:527

▲ *Terra Nova* in a gale. March 1912. Photographer: Herbert Ponting.
SPRI P2005/5/1167

One of his earliest decisions was that Campbell would lead the Eastern Party, which Scott intended to land at King Edward VII Land on the far side of the Great Ice Barrier.

While Scott assessed such issues, *Terra Nova* continued slowly towards Australia. On 12 October, she arrived at Melbourne, and suddenly Scott had other things to consider. Waiting for him was a puzzling message: 'Beg leave to inform you,' it stated, '*Fram* proceeding Antarctic. Amundsen.' It was clearly from the famed explorer Roald Amundsen, who several years before had been the first man to navigate the entire Northwest Passage. Scott was baffled by its meaning, as the Norwegian had been planning to conduct a drift through the Arctic basin in Nansen's old ship *Fram*. No one was certain what the enigmatic message could mean, and, for the time being, it remained a mystery.

On 29 November, *Terra Nova* sailed for the Antarctic from Port Chalmers, New Zealand. There was hardly any room aboard, particularly as 19 Manchurian ponies had been placed in specially constructed stables, and 33 Siberian sledge dogs had been widely spaced around the deck. These had all been brought to New Zealand by Cecil Meares, who was in charge of the expedition's dogs; Wilfred Bruce, Scott's brother-in-law; and two Russians: jockey Anton Omelchenko, who was to serve as the groom, and Dimitri Gerof, a dog-driver. Unfortunately, however, despite his expertise with dogs, Meares knew little about horses, which he had also been charged with selecting. To Oates' horror, almost all the ponies proved to be too old, diseased, lame or otherwise unsuitable for the work they would be expected to carry out.

Two days out of Port Chalmers, a violent gale hit the heavily laden ship. As huge waves broke over her, the main pump failed and all available men were set to bailing frantically. It was to no avail –

they could not keep up, and the ship was in clear danger of floundering. The only option was to repair the suction pump, but the hatch could not be opened. With desperate measures needed, a hole was cut through the engine-room bulkhead into a compartment filled with coal, some of the coal was moved, and another hole cut from there to the suction well. Teddy Evans squeezed past the remaining coal and down the pump shaft. There he managed to clear out the mixture of coal and oil that had choked the pump, and, with only a little time to spare, the ship was saved from a watery grave.

▲ Dr. Edward Wilson 21 April 1911. Photographer: Herbert Ponting. SPRI P2005/5/384

The difficulties with the pump were not the last of the problems, however. On 9 December, *Terra Nova* entered the pack ice at the northern reaches of the Ross Sea. It had taken *Discovery* only four days to clear through this region and *Nimrod* had done it even more quickly. This time the expedition was slowed for 20 days in the unremitting ice, not only draining coal supplies – limiting how long the ship could remain in the south – but setting back all the subsequent activities, including laying depots for the polar journey.

Yet another disappointment soon materialised. At Wilson's suggestion, Scott had considered using Cape Crozier, at the eastern end of Ross Island, as his base camp, rather than proceeding down McMurdo Sound. This would have three advantages: first, a lack of sea ice would not affect the summer depot parties; second, the expedition would have direct access to the Barrier at a point with fewer crevasses than the road south from McMurdo; and, third, Wilson would have the opportunity to study the unknown behaviours and incubation of the nearby emperor penguins, something that he had been keen to do since his *Discovery* days. However, the heavy swell at Cape Crozier made landing there impossible and Scott reluctantly turned the ship to McMurdo Sound.

Cape Evans

On 4 January 1911, with the way to his previous quarters at Hut Point blocked by ice, Scott decided to establish his base at a small promontory a dozen miles north of his original station. Called 'the Skuary' on the *Discovery* expedition, its name was now changed to Cape Evans, in honour of his second-in-command.

▲ Captain Robert Falcon Scott, 13 April 1911. Photographer: Herbert Ponting.
SPRI P2005/5/370

▲ Unloading the ill-fated motor-sledge. 8 Jan 1911. Photographer: Herbert Ponting.
SPRI P2005/5/161

▲ Organising the Cape Evans camp, 23 Jan 1911. Photographer: Herbert Ponting.
SPRI P2005/5/205

The dogs, ponies, two of the motor sledges and most of the prefabricated hut were unloaded immediately, and stores and equipment continued to be taken ashore during the next week. On 8 January, however, disaster struck when the third motor sledge plunged through the sea ice as it was being taken ashore. 'It's a big blow,' wrote Scott, 'to know that one of the two best motors, on which so much time and trouble have been spent, now lies at the bottom of the sea.'

The same week, another incident on the ice ended more happily, but only barely. Herbert Ponting, the talented but egocentric photographer who preferred to be called a 'camera artist,' went out onto the ice to photograph a pod of killer whales. Seeing his shadow, and perhaps thinking him to be a seal, the killer whales began their usual predatory behaviour, bumping the bottom of the ice floe in order to knock him into the water. The ice broke, but Ponting deftly leapt to the next floe, only to find himself pursued by the whales. As his colleagues looked on in horror, he managed to scramble from floe to floe ahead of the whales, finally reaching safety. 'What irony of fate to be eaten by a whale thinking one was a seal,' Campbell wrote wryly after Ponting's desperate escape, 'and then be spat out because one was only a photographer.'

As soon as the hut was ready, and before the summer waned and the darkness took over, four parties set out from Cape Evans. Scott left first, with 12 men, eight ponies and two dog teams, to lay depots on the Barrier in preparation for the all-important attempt on the Pole. Shortly afterwards, Taylor led the Western Party, consisting of Debenham, Wright and Petty Officers Evans and Robert Forde, to the mountains on the western side of McMurdo Sound for a geological survey. At the same time, Ponting, Day, Nelson and Lashly went north to Shackleton's old base at Cape Royds on a photographic excursion.

The other group was the six-man Eastern Party, commanded by Campbell and also including the geologist Priestley, the Naval Surgeon Murray Levick, Petty Officers George Abbott and Frank Browning, and Able Seaman Harry Dickason. However, they soon discovered that there was no way that *Terra Nova* could reach King Edward VII Land, so they reluctantly turned back west, hoping to find a landing spot on the Barrier itself.

On 3 February, the ship sailed into the inlet that Shackleton had named the Bay of Whales, where, to the company's shock, they discovered *Fram* moored against the ice edge. Amundsen, they soon learned, had arrived three weeks earlier and set up his base Framheim on the ice two miles inland – a full degree, or 60 geographical miles – closer to the Pole than Cape Evans. Amundsen was the picture of cordiality and visited *Terra Nova* before inviting Campbell, Pennell and Levick to join him for breakfast at Framheim. He even went so far as to offer to share his facilities with the British expedition. Needless to say, Campbell refused the offer, and he and Pennell sailed *Terra Nova* back

▲ The Eastern Party. Back row from left: Frank Browning, George Abbot and Harry Dickason. Front row from left: Victor Campbell and Raymond Priestley. Photographer: George Murray Levick. SPRI P97/185

to McMurdo to inform Scott of the Norwegian presence.

Scott had not yet returned from the depot-laying trip, so Campbell left a detailed message. *Terra Nova* then sailed north to find a location in which the newly re-designated Northern Party could winter. In mid-February, Campbell's party was left at Cape Adare, near the hut in which Carsten Borchgrevink's *Southern Cross* expedition had made the first wintering on the continent (1899). Little did Pennell know, as *Terra Nova* sailed for New Zealand, the terrible conditions that those remaining behind would experience in the next two years before they could extricate themselves.

Like Campbell's efforts, Scott's depot-laying had been filled with frustration. Although the dogs performed magnificently, the ponies on which he had counted so heavily had continual problems, demonstrating yet again their unsuitability for the Antarctic. Several early depots were established, but by the time Bluff Depot was set up at 79°S, the ponies were in very poor condition. Feeling they would be of little further use, Oates recommended taking them as far south as possible, then killing them and depoting the meat for the next spring. Scott, however, felt this was inhumane, and sent three of the weaker ponies back towards Hut Point. Shortly thereafter, distressed by the suffering of the remaining five beasts, he decided to leave the rest of the supplies at a point they named One Ton Depot, located at 79°29'S, some 30 miles short of their original goal. Again, Oates argued that the supplies should be taken as far south as possible, regardless of the

▲ Laying a depot. 8 February 1911. Photographer: Herbert Ponting. SPRI P2005/5/223

▲ Captain Oates and Petty Officer Abbott picketing the ponies on the sea ice. Photographer: Herbert Ponting. SPRI P2005/5/934

cost to the ponies. Again, Scott refused. Oates would later be proved tragically correct.

The return journey was a disaster, as six of the ponies were eventually lost. Perhaps even more upsetting for Scott was receiving the message from Campbell telling of the discovery of Amundsen and showing him beyond doubt that he was now in a race for the Pole. The lack of solid sea ice in McMurdo Sound meant he was now cut off from Cape Evans, and he and his party could do nothing but move into the old hut from his first expedition and wait for McMurdo Sound to freeze over again. Finally, on 13 April, only ten days before the Sun

▲ The 'Tenements' - bunks in the Cape Evans Hut.
Photographer: Herbert Ponting. SPRI P2005/5/538

▼ Herbert Ponting giving a lantern slide lecture on Japan, 16 October 1911.
Photographer: Herbert Ponting. SPRI P2005/5/1547

▲ Cherry-Garrard working on *The South Polar Times*, 8 June 1911. Photographer: Herbert Ponting. SPRI P2005/5/444

rose for the last time before winter, they were able to reach Cape Evans. There they were joined by Taylor's party, which had returned from the west.

Throughout the winter that followed, the hut at Cape Evans buzzed with activity. 'Sunny Jim' Simpson not only had meteorological equipment set up outside, but a special part of the hut was designated as the scientific workshop, where studies of zoology, parasitology, physics and geology were conducted. Immediately adjacent was the 6 x 8 foot (1.8 x 2.4 m) darkroom where Ponting not only carried out much of his work, but slept beside his photographic equipment and chemicals. In addition to the scientific programme, the members of the expedition prepared for the forthcoming sledging journeys and did those tasks required for daily existence in such a hostile environment. Entertainment and education were provided by Sunday services, regular musical performances, production of an expedition magazine, *The South Polar Times*, and a series of evening lectures.

The most important event of the period was undertaken by Wilson, Bowers and Cherry-Garrard, who made the first major sledging journey in the middle of the Antarctic winter, an excursion since immortalised as 'the worst journey in the world.' The winter journey was seen as a chance to gather sledging data in advance of the Pole journey, as well as being of great scientific interest. It would enable the men to try out different sledging rations under extreme field conditions, with varying quantities of fats and carbohydrates, to establish which would be the best rations for the Pole journey ahead. The expedition would also gather weather data from the Great Ice

Barrier in winter, which Simpson hoped might assist him in predicting weather conditions to be encountered on the Barrier during the attempt on the Pole.

It had long been Wilson's intention to visit Cape Crozier to study the incubation of emperor penguins and to collect their eggs. It was his hope that what were considered 'primitive' emperor penguin embryos would help establish the origin of birds and allow the testing of theories connecting their feathers to reptilian scales. Unable to establish the base at Cape Crozier, the only way to obtain the eggs was to travel to the penguin colony in the middle of winter, when the eggs would be laid.

▲ Henry 'Birdie' Bowers at Glacier Tongue. January 1911. Photographer: Herbert Ponting. SPRI P2005/5/1172

▼ Sledging ration for one man for one day. Photographer: Herbert Ponting. SPRI P2005/5/1393

▲ Bowers, Wilson and Cherry-Garrard beside their sledge, setting out on their winter journey to Cape Crozierb27 June 1911.
Photographer: Herbert Ponting. SPRI P2005/5/452

On the morning of 27 June, the three set off in the dark on the 57-mile (105-km) trip, hauling food and equipment on two sledges totalling 757 pounds (341 kg). They could not cross Ross Island because Mount Erebus stood in the way, so they went south past Hut Point, ascending onto the Great Ice Barrier, where the temperatures soon dropped to –49°C (–56°F). The rough surface and sand-like snow on the Barrier made hauling the two sledges at the same time impossible, so they were forced to relay, first hauling one sledge and then returning for the other, meaning they gained only one mile for each three travelled. It was brutally hard work, made worse by the fact that they had to stumble ahead blindly over pressure ridges and crevasses, where one false move in the darkness might end in death. Each night they shivered inside sleeping bags woefully inadequate to protect them from the cold, while ice formed inside and outside the bags.

▲ Wilson, Bowers and Cherry-Garrard on return from winter trip to Cape Crozier. 1 August 1911. Photographer: Herbert Ponting. SPRI P2005/5/466

▼ *An Ice Berg off C. Evans. Sept. 1.11. 4.30pm.* Watercolour by E.A. Wilson. SPRI Museum N: 1541

On 5 July, the temperatures plummeted even further, reaching a low of −61°C (−77°F), but the three continued on, and ten days later they arrived at what they called 'The Knoll' on a slope above the colony. There they built a stone 'igloo', with a piece of canvas for a roof. On 20 July, after one failed attempt, they were able to complete a tortuous descent to the colony, where, despite the disappointment at how few penguins there were, they caught and killed three to use the blubber for their stove, and collected five eggs, two of which were broken on the climb back to the igloo. Almost immediately, a gale prevented them from returning to the colony, blew away their tent and destroyed their canvas roof, exposing them to the worst of the conditions.

After the gale died down, they began what must have seemed a hopeless search for the tent, without which they would certainly perish on the return journey. Remarkably, Bowers found it about a quarter of a mile away. They began their return on 25 July and, despite the journey being no easier than on the way out, managed to reach Hut Point on the last day of the month. Pushing ahead as fast as they could, they arrived the next night at Cape Evans, where they had to be pulled forcefully from their frozen clothes. Their journey had taken them 114 geographical miles (210 km) in 36 days, but they had been perhaps the harshest, most demanding days in the history of Antarctic exploration. Ponting took a famous picture of the three men around the table immediately after their return, about which he wrote: 'their faces bore unmistakable evidence of the terrible hardships they had endured. Their looks haunted me for days.'

▲ The three Emperor penguin eggs collected from Cape Crozier, now in the Natural History Museum. Photographer: Herbert Ponting. SPRI P2005/5/840

The Southern Journey

The great effort that Scott had been looking forward to for years finally got underway on 24 October, when Teddy Evans left base in charge of the two motor sledges, each of which pulled behind it three sledges loaded with food, fuel, and equipment. The plan was for the sledges to proceed past One Ton Depot and then due south to 80°30'S, where they would rendezvous with the pony party.

This group consisted of Scott and nine others, each leading a pony, all of whom left Cape Evans on 1 November. The dogs were also involved, but the superior speed and efficiency they had already demonstrated led Scott to order Meares and Dimitri to follow later, with 23 dogs hauling two sledges.

Sadly, it was not long before Scott's carefully planned procession began to fail. On 4 November, before the pony party reached Corner Camp, one of the first small stations established, they passed a motor sledge that had broken down and been abandoned. Two days later, the remains of the second motor sledge marked the end of the venture into modern technology that had cost Scott so much

▲ Captain Scott and the Southern Party. Mount Erebus in background. 26 January 1911. Photographer: Herbert Ponting. SPRI P2005/5/215

▲ The motor sledge party. From left: Lashly, Day, Evans, Hooper. October 1911. Photographer: Herbert Ponting. SPRI P2005/5/574

financially. 'We both damned the motors,' Oates noted after talking to Meares about their failure. '3 motors at £1000 each, 19 ponies at £5 each, 32 dogs at 30/– each. If Scott fails to get to the Pole, he jolly well deserves it.' Meanwhile, Evans and his party – Day, Lashly and the steward, F.J. Hooper – had continued on with as much as they could take by man-hauling.

On 21 November, the three elements of the southern party were re-united at the rendezvous station, which became known as the Mount Hooper Depot in honour of the steward, who had helped build a monstrous cairn while waiting for the other groups. With the different parties now leaving from the same camp, five separate starts were made daily, spreading the men and animals across the Barrier. Evans' man-hauling team left two to three hours before those with the ponies, who departed in three teams, based on the animals' condition. Several hours after the last of these, Meares, Dimitri and the vibrant dog teams would race off, showing time after time that they were far and away the most efficient form of transport.

After three days of this pattern, the wear on the ponies was becoming evident and the first was shot. Its leader, Edward Atkinson, the surgeon, joined Evans' team, allowing Day and Hooper to return to Cape Evans with a sledge and two sick dogs. Another two days' efforts allowed the establishment of the Mid-Barrier Depot at 81°35'S. Soon thereafter, another pony was put down and Wright, the physicist, joined the man-hauling team. The 14 men still on the Barrier must have received an emotional boost not long thereafter when they passed Scott's original farthest south, set almost nine years before.

By early December, five of the ten ponies had been shot, and heavy snow was slowing the others to a crawl. Nevertheless, the party was on the verge of reaching the Beardmore Glacier when a warm blizzard blew in with such ferocity that they were tent-bound for four days, while pools of water formed under and around the tents, soaking everyone. When the gale finally eased, the party moved forward again, but the fact that the ponies had been forced to remain more or less inert during the preceding terrible conditions had done them in, and that night, at what was named 'Shambles Camp,' the remainder were killed.

On 10 December, at the foot of the Beardmore, the party was reorganised into three man-hauling teams, which comprised Scott, Wilson, Oates and Edgar Evans; Teddy Evans, Atkinson, Wright and Lashly; and Bowers, Cherry-Garrard, Crean and Petty Officer Patrick Keohane. The next day, after helping establish the Lower Glacier Depot, Mears and Dimitri raced away north with the dogs, emphasising in their way the advantages that Amundsen would have over the British man-hauling teams.

The men now ploughed slowly ahead, slogging their way up a ceaseless grade with between 700 and 800 pounds (330–360 kg) per sledge. Evans' group quickly began to show signs of exhaustion, as they had been man-hauling while the others had been leading ponies. The route up the Beardmore proved to be some 105 geographical miles (195 km), rising to an elevation of more than 10,000 feet (3050 m). Progress was slow and brutally difficult, and Scott regularly began to voice concern in his diary. 'We must push on all we can, for we are now 6 days behind Shackleton,' he wrote, making one of his frequent comparisons to the rate travelled by his great rival.

Scott should perhaps have been more concerned with the progress of Amundsen, but little did he know that the Norwegians were already far ahead of the British team. In fact, while Scott was kept in his tent by the long blizzard, Amundsen had surpassed Shackleton's farthest south, putting him within 97 geographical miles (180 km) of the Pole. On 14 December, Amundsen and his team – Olav Bjaaland, Helmer Hanssen, Sverre Hassel and Oscar Wisting – became the first men ever

▲ Amundsen and his men at their South Pole camp, Polheim, 14 December 1911. Photographer: Olav Bjaaland. Courtesy of National Library of Norway.

to reach the South Pole. Three days later, they left the tent they had named Polheim to begin their return north, having taken the prize.

Meanwhile, the British continued their slow, grinding ascent of the Beardmore, and on 21 December they established the Upper Glacier Depot, at 85°7'S. The next morning the four men Scott considered the weakest were sent back: Atkinson, Cherry-Garrard, Wright and Keohane. Actually, Oates was as in poor shape as any of them, due to problems with an old wound he had received in his thigh and with his feet, about which he wrote: 'They have been continually wet since leaving Hut Point and now walking along this hard ice in frozen crampons has made rather hay of them.' Atkinson, a close friend of Oates, later told Cherry-Garrard that 'Soldier' had not wanted to continue, but, in a time when physical weakness was seen as something to be ashamed of, he maintained a stoic exterior and refused to reveal his thoughts to Scott.

Atkinson took back with him an important change of plans. Scott had increasing concerns about the return across the Barrier, and decided that the dogs could aid substantially in the process. However, he and Meares had quarrelled, with the latter indicating that he was going north on *Terra Nova*, which would probably sail before Scott's return. Therefore, Scott instructed Atkinson to bring the dogs south later in the season to meet the Polar Party.

Meanwhile, the two remaining sledging parties – Teddy Evans' now consisting of himself, Bowers, Crean and Lashly – pounded on, still going uphill for two more weeks, and gradually wearing down most of the men. Scott, who was almost certainly the most powerful of the entire group, perhaps did not understand the physical or mental limits of some of the others, who were more obviously affected than he was by the lack of caloric intake, dehydration, low temperatures and high altitude. Instead, he seemed to focus on the problems of Evans' team, and on 31 December he ordered them to depot their skis and continue on foot. This remains one of the strangest decisions of the expedition, because Scott knew that it was easier to travel on skis. In fact, two days later he wrote, 'It's been a plod for the foot people and pretty easy going for us.'

Most of the 'foot people' did not have to deal with Scott's decision for long, because on 3 January 1912, he told Evans, Crean and Lashly that they would be returning to base. However, in a remarkable change of plans, he asked Evans to allow Bowers to join the Polar Party. The request was little more than a formality, of course – as a subordinate, Evans could not refuse. Scott's decision to take five men to the Pole has never been fully understood, and has been controversial for decades. Most importantly, it undermined the logistical planning that had hitherto been so carefully followed. It meant that Evans' returning party would be undermanned, and it left Scott with an extra man in an already small tent. Although Scott may have believed that a fifth man hauling the sledge would outweigh any difficulties, the decision might have been more impulsive than considered, particularly as Bowers was one of the men whom Scott had ordered to cache his skis.

So how *did* Scott make his determination of the final Polar Party? There had never been any doubt that Wilson – Scott's confidant, trusted advisor and proven sledging companion from *Discovery* – would be selected for the final push to the Pole. Bowers, although not initially one of Scott's coterie, had proven himself invaluable as a navigator, an organiser of the stores and a tireless, unquestioning worker. The selections of Oates and Edgar Evans, however, were, from an outside standpoint, a little less understandable.

It has been suggested that the choice of Oates was in part dictated by Scott wanting a representative of the other great service – the army – to join him at the Pole. Whether or not this is accurate, Oates' work with the ponies had certainly proven his value, although taking care of them each night meant that he worked for longer periods than most of the others, and therefore had missed a number of meals. Moreover, during the ascent of the Beardmore, he had suffered increasing problems with his feet and his old war wound. Nor was his desire to attain the Pole all-encompassing like that of several of his comrades, although he did feel the need to continue for the honour of the army. Thus, Oates would certainly give the final journey everything he had, but the problem was that much of his strength, stamina and energy had already been expended.

Similarly, Scott might have selected Edgar Evans in part wanting the lower deck represented at the Pole. Scott also had a high personal regard for the Welshman, who had played a major role in Scott's ascent to the Plateau through the western mountains on the *Discovery* Expedition. Scott perceived Evans as the epitome of strength, a powerful man whose very physique gave confidence in his abilities. However, although Evans also had utter devotion to his leader, the increase in his drinking through the years had taken a toll, and Wilson and Atkinson had agreed that Lashly – a teetotal non-smoker – was far the best choice of the seamen, while Crean would clearly have been a better selection than Evans.

As the Polar Party continued south, the problems caused by the change of plan began to become apparent. Bowers soon found himself perpetually exhausted, as he was forced to keep up with companions moving easily on skis. Moreover, it took considerably longer to cook for five than for four, reducing the supply of fuel at an alarming rate. Evans also suffered continuing problems with his hand, which had been badly cut on 31 December when he and Crean had been changing sledge runners.

For the moment, however, these issues could easily be overlooked, as the party drew steadily closer to the great goal. Even the return journey did not seem so daunting, as Scott had sent verbal orders back with Teddy Evans, which would help make for a safe return. His new plan was for Meares to bring the dogs to between 82° and 83°S – much farther south than previously planned – in the middle of February. Not only would this guarantee the Polar Party extra supplies, it would allow the news of the Pole to reach *Terra Nova* before she departed for New Zealand. Unfortunately, this directive would never be acted upon, and that failure would have dire consequences for the five men still heading south.

A Tragedy in the Making

Despite Scott's confident plans, Evans, Crean and Lashly struggled mightily on their trek hundreds of miles back to Cape Evans. Nothing they had yet faced prepared them for the nightmare journey down the Beardmore Glacier.

Even Lashly, as tough a man as can be imagined, noted: 'I cannot describe the maze we got into and the hairbreadth escapes we have had ... The more we tried to get clear the worse the pressure got; at times it seemed almost impossible for us to get along.'

In the effort to find a safe path, Evans removed his goggles to see better and, as a result, he became snow-blind. Now he was unable to pull the sledge through the most difficult areas, thereby condemning Crean and Lashly to hellish hardships. Then, in late January, as they left the Beardmore for the Barrier, they realised that Evans was suffering from advanced scurvy. He became progressively weaker and in mid-February he suddenly passed out in the middle of a march, leading Crean and Lashly initially to assume that he had dropped dead.

▲ The Polar Party man hauling sledge. Photographer: Henry Bowers. SPRI P48/281/1A

Evans was actually still alive, but in no condition to do any work. Despite their own exhausted condition, Crean and Lashly would have to pull him back the rest of the way, so they discarded everything they no longer needed to survive and tenderly strapped Evans onto the sledge. He pleaded to be left behind and not burden them, but the two men simply continued on their march, hauling him day after day, all the way across the Barrier. Reaching the southernmost of the motor sledges raised their spirits, but the next morning, Evans was too dangerously ill even to be moved. With no other option, Lashly stayed behind to care for him, while Crean staggered on alone to find help. His solo trek took him more than 30 miles (55 km) to Hut Point, where, at 3:30 on the morning of 19 February, he found Atkinson and Dimitri with the dog teams, which they had brought out to restock One Ton Depot.

Evans was not out of the woods yet, however. Atkinson could not go to him immediately, because a sudden fog made driving the dogs too unsafe. But when the conditions cleared, Atkinson and Dimitri raced out onto the Barrier to save the sick man and also to bring back Lashly. Had the timing been off by a day or two, Evans would surely have died, but as it was, the surgeon was able to give him constant attention in the following weeks, ensuring his survival. Lost in the shuffle, however, was that Evans' desperate condition prevented him from passing along Scott's latest message about the dogs meeting the Polar Party. When *Terra Nova* arrived, Evans, still critically ill, was taken back to New Zealand to recover. Crean and Lashly were each later awarded the Albert Medal for their heroic actions.

The general feeling at Cape Evans was buoyed by Evans' rescue, and even more so by the reports of Crean and Lashly that all the members of the Polar Party had been strong and healthy when they left. It was only Tryggve Gran – the Norwegian ski instructor who had been recommended by Nansen, and who had a better feel for snow and ice than most of the Britons – who believed that the joyful anticipation was misguided. Instead, Evans' condition told Gran a frightening story. 'My conversation with Evans had not lasted long, but from what I heard,' he wrote, 'the prospects of our five-man polar party were not so bright as most of the members of the expedition imagined. Evans' frightful return journey was a pointer to what Scott and his men would be bound to undergo.'

The Pole Without Priority

Before Evans and his two saviours-to-be had even left the Beardmore, Scott's party had achieved the highest and lowest of emotional states. On 9 January 1912, they surpassed the mark of 88°23'S that Shackleton had established three years to the day before.

At that time, Shackleton had dearly wished to continue to the Pole, but had understood innately that the members of his party had gone as far as they safely could. Perhaps the greatest achievement of Shackleton's career had been having the courage to turn back despite being so close to his goal. Yet even then, the four men had gone through incredible struggles in order to return alive – and had barely done so. On the same day of the year, Scott and his companions – now cold, exhausted, undernourished, dehydrated and undoubtedly suffering from the early stages of scurvy, but buoyed up by having established what they believed to be a record latitude – ignored Shackleton's lesson and continued south.

▲ The Polar Party at Amundsen's tent, photographed by Henry Bowers, 18 January 1912. SPRI P48/281/4

▲ Edward Wilson's sketches of Amundsen's Pole markers. SPRI N:546

A week later, their spirits crashed when they saw a strange feature ahead. As they approached it, according to Scott, it turned out to be 'a black flag tied to a sledge bearer; near by the remains of a camp; sledge tracks and ski tracks going and coming and the clear trace of dogs' paws – many dogs. This told us the whole story. The Norwegians have forestalled us and are first at the Pole.'

Each man must have been heartbroken in his own way, a feeling made worse on 17 January, when they approached the Pole itself. 'Great God! this is an awful place,' wrote Scott, 'and terrible enough for us to have laboured to it without the reward of priority.' The following day, they discovered a tent the Norwegians had left behind. Inside were the names of the five men who had beaten them, an assortment of items they did not need on their return journey, and a letter to the King of Norway, with a note to Scott asking him to forward it to King Haakon VII should Amundsen not return safely. The devastated British explorers took pictures of themselves at the Pole and dejectedly headed north. 'Now for the run home and a desperate struggle,' Scott wrote. 'I wonder if we can do it.'

Initially, there was no reason to think that they would not. They managed to make regular progress, helped at times by rigging a sail on the sledge, which was propelled ahead by a wind from the south. On the last day of the month they found Bowers' skis, and he could finally move ahead as easily as his comrades, but any such assistance was offset by the effects of scurvy, malnutrition, constant cold and depression. Moreover, nagging injuries made travel more and more difficult. Oates' toes turned black from frostbite; Evans' hands became

▲ *Sledge sailing in a blizzard*, by Edward Wilson. SPRI Museum N:1400

raw and sore, his fingernails began to fall off and his month-old cut would not heal; and Wilson had trouble walking after he strained a tendon in his leg.

On 7 February, their arrival at the Upper Glacier Depot signalled that they were off the Plateau and heading down the Beardmore. The spirits of Wilson and Scott rose as they sensed that conditions would soon improve, but Evans' emotional state declined, as did his physical ability. Scott and Wilson felt confident enough about their situation that parts of each of the next two days were spent obtaining geological samples, which later proved to be of significant scientific value.

But on 11 February they became hopelessly lost in maze of crevasses part way down the glacier. The could not find a path ahead even by the next day, when, according to Wilson, they wandered 'absolutely lost for hours and hours.' That evening, they found their supplies had shrunk to one meal. Desperate, they received a reprieve the next day, when Wilson spied the Mid-Glacier Depot. Unfortunately, even that only provided three and a half days' food.

Evans had by now slipped into serious mental and physical decline. 'Evans has nearly broken down in brain, we think,' Scott wrote on 16 February. 'He is absolutely changed from his normal self-reliant self.' But the next day his condition took another turn for the worse. He was taken out his harness in order to follow the others, but in even this he proved incapable, and he had soon dropped far behind.

When his companions returned to find him, he was totally disoriented, on his knees and with his hands uncovered. The others put him on a sledge and brought him to camp, but he quickly became comatose and died that night.

At that moment, Evans' death – as hard as it would have been on his comrades – must have seemed a blessing in some senses, as he would have continued to impede any progress. 'It is a terrible thing to lose a companion in this way,' Scott wrote, 'but calm reflection shows that there could not have been a better ending to the terrible anxieties of the past week.' The next day they left the glacier behind, but their hopes for better conditions were soon dashed, as they found the Barrier surface so difficult to pull the sledge on that they sometimes only achieved five geographical miles (9.3 km) in a day. As if these conditions were not bad enough, the temperatures suddenly dropped fiercely, reaching −40°C (−40°F) on 27 February.

On 1 March, the excitement of reaching the Middle Barrier Depot was tempered by the discovery that there was considerably less oil there than expected, meaning they would be fortunate to eke it out until the next depot. Worse yet, after more than two months of horrific pain and struggle, Oates revealed the degenerating state of his extremely frostbitten feet. They were not his only problem. Vitamin C is needed to keep scar tissue together, and it has been speculated that, accompanying the scurvy, the scar on his thigh from his war wound would have begun to dissolve and thus reopen. The other members of the party were also undoubtedly suffering from the advanced stages of scurvy.

Despite such adversity, the party still moved with incredible determination across the seemingly endless Barrier. Each man was growing weaker,

▲ Wooden sundial with metal gnomon, made by Raymond Priestley, and used by Apsley Cherry-Garrard, *Terra Nova* expedition, 1910-13.
SPRI Museum Y:66/13/17

particularly Oates, who was soon taken out of harness, as he was no longer able to pull the sledge. Finally, on 9 March the party trudged into the Mount Hooper Depot, but, to their great dismay, it had not been restocked. 'Shortage on our allowance all round,' Scott wrote. 'The dogs which would have been our salvation have evidently failed.' In fact, due to Teddy Evans' condition, Scott's latest plan had never been mentioned, and those left at Cape Evans followed his earlier orders, which had forbidden risking the dogs on a lengthy journey. Cherry-Garrard, who had taken over driving the animals in the absence of Meares and then Atkinson – who would not abandon Evans – and then Wright, who would not leave his scientific studies, tortured himself about going farther south from One Ton Depot. But for days he and Dimitri were trapped in atrocious conditions by a blizzard and when at last it blew out, he was determined to follow Scott's orders to the letter, so he returned to base without being able to help the desperate Polar Party.

For Oates, meanwhile, every moment must have been intense agony and, on 15 March, he asked to be left behind in his sleeping bag. The others would not, of course, accommodate such a request, and, with a heroic effort, he managed another few miles, but the next morning, when he woke, 'It was blowing a blizzard,' wrote Scott. 'He said, "I am just going outside and may be some time." He went out into the blizzard, and we have not seen him since ... We knew that poor Oates was walking to his death, but though we tried to dissuade him, we knew it was the act of a brave man and an English gentleman. We all hope to meet the end with a similar spirit, and assuredly the end is not far.'

It could not be otherwise. All three men – Scott, Wilson and Bowers – were failing physically. Within two days of Oates' sacrifice, Scott, one of the most powerful, determined man-haulers ever, had become a shadow of his former self, as his right foot became severely frostbitten. Nevertheless, the three bravely staggered on to a point only 11 geographical miles (20 km) south of One Ton Depot – tragically, well north of the place where, the previous autumn, Oates had argued the depot should be established despite the cost to the ponies.

With their desperately needed supplies so near to hand, the men's progress was halted as a severe blizzard blew in, trapping them in their tent. For the next few days, they planned to make a dash for One Ton Depot as soon as the gale abated, but cease it would not, and, as the

▲ Three men in a pyramid tent, pencil sketch by Edward Wilson. SPRI N:1396

three already weakened men ran out of oil, food and then physical strength, all they could do was await the end. Before that came, Scott wrote a series of letters and a message to the public, enumerating the causes for the tragedy that had befallen the Polar Party and justifying his decisions and his command. In these, the rich texture of Scott's language and his emotive messages eventually helped make him one of Britain's most beloved and iconic heroes. 'Had we lived,' he wrote, 'I should have had a tale to tell of the hardihood, endurance and courage of my companions which would have stirred the heart of every Englishman. These rough notes and our dead bodies must tell the tale ...'

And then, all that was left was the interminable blizzard outside the tent.

The Last Discovery

It was a sad second winter at Cape Evans, as the members of the Polar Party had to be assumed to be dead, those of the Northern Party had not been rescued, and a number of the key figures from the first year – among them Simpson, Taylor and Ponting – had sailed north on *Terra Nova.*

As senior officer, Atkinson assumed command, and he ran the base efficiently. During this time, the 13 men were united by an already established camaraderie and by the continuing scientific efforts of Debenham, Wright, and Nelson.

On Midwinter Day, Atkinson gave the entire party the chance to determine the priorities of the coming spring: either attempting to discover the fate of Scott and his men or trying to relieve the Northern Party. However, such were the difficulties in reaching the latter, and the strength of the hope that *Terra Nova* would be able to relieve them this year, that it was unanimously agreed to try to discover what had happened to Scott. Yet it was not an easy choice.

▲ Dr Atkinson in the laboratory. 15 September 1911.
Photographer: Herbert Ponting. SPRI P2005/5/492

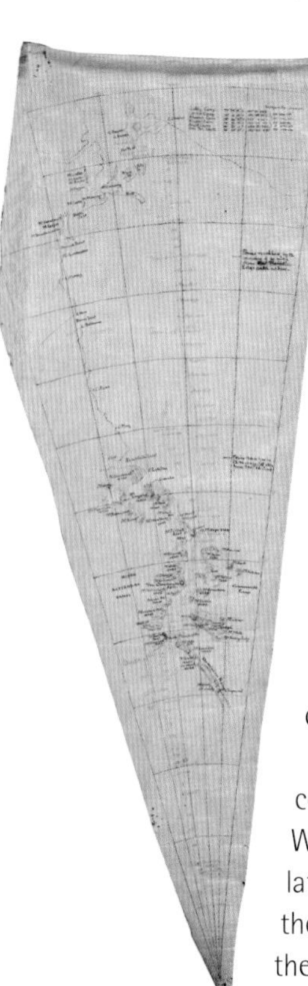

◀ Tracing of map of the route from Ross Island to the South Pole prepared by Dr Edward Wilson and found in the tent with his body. SPRI Museum N:1036

Cherry-Garrard later wrote that, 'it seemed to me unthinkable that we should leave live men to search for those who were dead.'

Nevertheless, the course of action was set, and Atkinson's plan sent out 11 men in three parties, three with the dogs, which would travel to the base of the Beardmore Glacier, and two four-man groups guiding the seven mules that, at Oates' recommendation, had been brought south the previous autumn. The mule parties were to ascend the Beardmore, one of them continuing all the way to the top if necessary.

On 30 October, the mule parties left under the command of Wright, with Nelson, Gran, Lashly, Crean, Williamson, Keohane and Hooper joining him. Two days later, Atkinson, Cherry-Garrard and Dimitri followed with the dog teams. As always, the dogs soon caught up with the slower moving animals, and from there on they travelled together. On the morning of 12 November, approximately 11 miles (20 km) beyond One Ton Depot, Wright saw what he thought was a cairn and veered off towards it. It was the cap of a tent.

The snow was brushed off and Atkinson and Lashly entered. There, in their sleeping bags, lay Wilson, Scott and Bowers. With them were their diaries, meteorological log and letters. Scott had also left a request that whoever found them should read his diary and take it home. After Atkinson did so, he gathered his party

▶ Charles Wright working with his transit. 8 August 1911. Photographer: Herbert Ponting. SPRI P2005/5/470

around and read the account of Oates' death and Scott's 'Message to the Public.' Having removed the letters and other personal items to return to the families, the members of the search party then collapsed the tent over their fallen comrades, and built a cairn on which they placed a cross in their memory.

After conducting an unsuccessful search for Oates' body – and finding only his sleeping bag – Atkinson erected another cairn with a small cross near where Oates' act of extraordinary bravery and sacrifice had been carried out. That done, attention returned to the assistance of the Northern Party. The dog teams quickly pushed ahead of the mules, and on 25 November Atkinson, Cherry-Garrard and Dimitri reached Hut Point. 'Cherry went into the hut,' Atkinson wrote, 'and returned with a letter and his face transformed.' All of the members Northern Party, the letter told them, were safe.

▲ Captain Oates' reindeer skin sleeping bag. SPRI Museum N:857

Two months later, on 18 January 1913, *Terra Nova*, now under the command of a fully recovered Teddy Evans, arrived at Cape Evans. It became a time of sadness, when the shipboard members of the expedition joined the shore party in mourning the tragedy of the Polar Party, particularly as they already knew of Amundsen's success. At the same time, they celebrated the triumph of the members of Scott's party, who had, as Scott had written, 'shown that Englishmen can endure hardships, help one another, and meet death with as great a fortitude as ever in the past.' They had, in fact, demonstrated those magical, enduring qualities that had made the British Empire the greatest in the world.

▲ The cairn raised over the final camp. SPRI P2005/5/1249

Two days later, Atkinson led a group to the top of Observation Hill, which overlooks both Hut Point and the Great Ice Barrier. There they erected a nine foot (2.7 m) memorial cross to the Polar Party: Scott, Wilson, Bowers, Oates and Evans. It still stands there today, honouring the valiant struggle and achievements – the triumph, failure, and tragedy – of that brave collection of men.

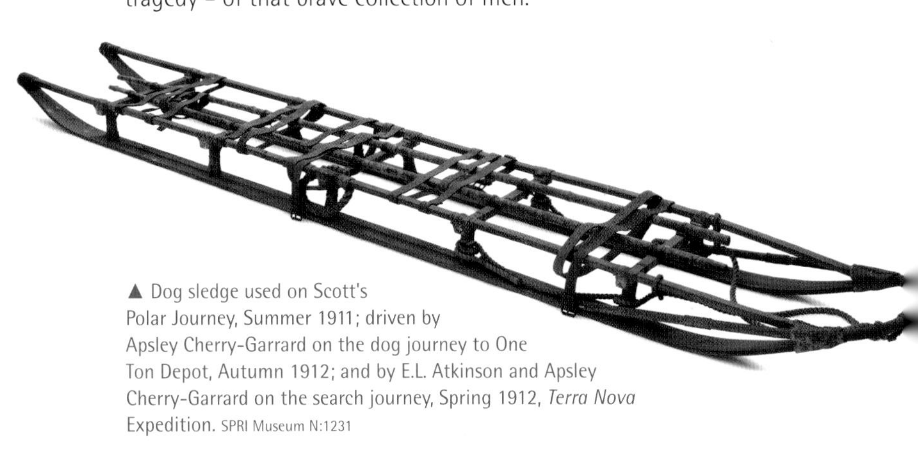

▲ Dog sledge used on Scott's Polar Journey, Summer 1911; driven by Apsley Cherry-Garrard on the dog journey to One Ton Depot, Autumn 1912; and by E.L. Atkinson and Apsley Cherry-Garrard on the search journey, Spring 1912, *Terra Nova* Expedition. SPRI Museum N:1231

The Northern Party

If the tragic tale of Captain Scott's Polar Party 'would have stirred the heart of every Englishman,' one can only imagine how the terrible drama of the Northern Party of the *Terra Nova* expedition must have frightened, horrified, and gripped in equal measure the parts of that same public that heard the story.

Although the achievements of the Northern Party were little noticed by some in the crushing grief and sense of national bereavement that surrounded the death of Scott and his companions, the reality was that the six men of the Northern Party faced seemingly equally overwhelming odds against their survival, but all somehow lived to tell the tale.

The Northern Party came into existence when the Eastern Party failed to reach King Edward VII Land or to find a location to land on the Great Ice Barrier other than that already occupied by Roald Amundsen's expedition. After leaving a message for Scott about Amundsen, Victor Campbell and Harry Pennell took *Terra Nova* north, following Scott's instructions about using the region of Robertson Bay as an alternative site, while also attempting to allow the ship to escape the new-forming ice.

The logical area for establishing a base along the northern coast was actually well to the west of Robertson Bay, as no significant exploration had been conducted beyond Cape North. However, it proved impossible to land there and, worried by the lack of coal and increasingly poor weather, the party turned about-face and returned to Cape Adare, where, on 18 February 1911, they started unloading at the site at which Carsten Borchgrevink and the members of his *Southern Cross* Expedition had carried out the first wintering on the Antarctic continent. Although some geographical and scientific work had therefore previously been conducted in the vicinity, that to be carried out by the Northern Party – tentatively similar to what had been intended for King Edward VII Land – promised initially to be the most comprehensive to date in that region of the Ross Sea.

Immediately, a small hut was built, and Borchgrevink's battered huts restored. When Pennell sailed north on 20 February, he left behind Campbell and a party consisting of Naval Surgeon Murray Levick;

◀ Ice axe used by Raymond Priestley to dig out the snow cave in which the Northern Party wintered. SPRI Museum Y:74/8

geologist Raymond Priestley, who was afforded officer status; Petty Officers George Abbott and Frank Browning; and Able Seaman Harry Dickason. The party quickly settled into a scientific routine, including making regular geological, zoological and magnetic observations, in addition to round-the-clock meteorological measurements. Although not trained for the purposes of science, and not as helpful in those areas, Abbott and Dickason proved invaluable, with the former being a talented carpenter and the latter showing himself to be an excellent cook and baker.

It soon became apparent that Cape Adare was a disappointing location from which to conduct geographical exploration or, in fact, scientific research. At the beginning of May, the winter was announced by a fast and furious blizzard. Throughout the long darkness that followed, Campbell – nicknamed the 'wicked mate' for his obdurate manner and foul temper – required unyielding social and hierarchical structure, and maintained strict order with firm discipline and an uncompromising routine.

Campbell's original plan was for the six men to travel in two exploring parties across Robertson Bay – the long basin separating the finger jutting north that was Cape Adare from the unknown regions to the west. However, early journeys demonstrated that the heavy salt content in the bay ice made sledging too slow for great distances to be feasible. These problems were compounded in the middle of August, when a gale blew much of the ice out of the bay, requiring a much longer route along the coast. Even attempting this supposedly safer route, Campbell was stopped by thin ice and had to retreat.

Due to the breakup of the ice, for much of their time at Cape Adare, the men were essentially trapped near their base and did little but the most basic of scientific tasks. Then, suddenly, on 3 January 1912, *Terra Nova* appeared, with Pennell demanding an immediate withdrawal from the site, due to his concerns about the pack ice, the lack of soundings near the shore and the makings of a gale coming in behind them. To their horror, the six members of the shore party were required to board without even being given the time to load their scientific samples.

In order to achieve some of the unaccomplished goals, Campbell convinced Pennell to leave the party near Evans Coves in *Terra Nova* Bay, farther along the Victoria Land coast. They agreed that *Terra Nova* would return for them six weeks later, after collecting the members of the Second Western Party – another geological survey under Griffith Taylor. But Campbell's impatience and Pennell's cautiousness jointly led to a disastrous decision. 'We had prepared a large depot at Cape Adare which was to have been landed with us here,' Priestley later wrote, 'but it was necessary to sledge all our gear about half a mile over sea ice before it would have been possible to depot it, and as Campbell did not wish to delay the ship, he decided to land only such spare food as could be taken in one journey by ourselves and a sledge party from the ship's crew.'

▲ Levick and Campbell skinning and gutting a seal. Frank Debenham Collection. SPRI P54/16/386g

That spare food consisted only of six weeks' sledging rations plus a month's skeleton rations and a small amount of other emergency fare – the last two parts in case *Terra Nova* could not return by its set date of 18 February. The sledgers immediately began to survey the surrounding region and to compile a variety of scientific collections, including segments of fossilized tree trunks, which later provided early sources of information on past Antarctic climatic conditions.

In early February, the men returned to await the ship, but, as the days went by, they became more and more worried. In fact, Pennell made several gallant efforts to reach them, but each proved futile because of exceptionally heavy ice, and finally the ship was forced to sail to Cape Evans, where Pennell found just the opposite – a lack of ice in McMurdo Sound had cut Cape Evans off from Hut Point. At the base, surgeon Edward Atkinson took aboard Teddy Evans – who was deathly ill with scurvy – to be shipped to New Zealand as a patient. *Terra Nova* then made another unsuccessful attempt to reach Campbell's party, and then, his lack of coal dictating a quick trip home, Pennell headed to New Zealand, effectively leaving Campbell's party to its fate.

After an early attempt to search for Scott's Polar Party failed, Atkinson turned his attention to Campbell and his men. With winter nearing, Atkinson, physicist Charles Wright, and Petty Officers Patrick Keohane and Thomas Williamson marched north along the coastline, but could find no trace of their comrades, and therefore grudgingly returned to Hut Point for a winter full of uncertainty about the Northern Party's fate.

Meanwhile, Campbell and his men faced the prospect of a winter without shelter, heat or adequate food. However, with indefatigable spirit, they began excavating an underground cave, which was hacked from the ice and snow that covered a granite outcrop they named Inexpressible Island. Known as 'the igloo,' the tiny dwelling had a set of steps leading down from a doorway that could be closed off by a sealskin curtain. To one side was a latrine, which became hellish as all the men suffered from a lack of fibre and carbohydrate. The small passageway included two doorframes made from biscuit cases, and fitted to keep out the wind and help warm the interior.

The inner room – measuring 12 x 9 feet (3.6 x 2.7 m), with a ceiling so low no one could stand upright – was where all six men slept, cooked, ate and spent their waking hours. Remarkably, Campbell's martinet style made the room even smaller in some ways. Early on, he drew a line with his boot down the centre of the floor. Pointing to one side of the cave, he designated it the mess-deck, where the three seamen would sleep; the other side, on which the two officers and the scientist slept, he indicated was the quarterdeck. 'As on board ship,' he informed them, 'everything that is said and done on the mess-deck will be the responsibility of the men and it shall not be heard or paid attention to or interfered with by any of the officers who reside on the quarterdeck. And the opposite is true.'

Under these conditions, particularly through the seemingly endless, dark, cold winter, food was all-important. However, with very few rations left, most of their diet was seal and penguin, the killing of which also provided blubber for their small cooking stove. The meat

▲ Northern party on arrival at Cape Evans, 7 November 1912. From left: Dickason, Abbott, Browning, Campbell, Priestly and Levick. Frank Debenham Collection. SPRI P54/16/7

and blubber were initially accompanied by a biscuit per day, but this decreased dramatically in August. This diet brought about numerous physical problems, and Browning in particular almost died from the complications. A highlight of the winter came when Browning killed and cut open a seal to find 36 edible fish inside. Thereafter, the spying of a seal was announced with the cry of 'Fish!!'

Even the procedure of cooking such small quantities led to problems. If placed too close to the wall, the stove caused the ice sides to melt. And the blubber produced a greasy, choking pall that caused the men's eyes to smart. It also saturated their clothes with a combination of smoke and oil that Browning dubbed 'smitch.' This horrible substance ruined the fur in the sleeping bags, causing them to lose their warmth, and it made gloves and mitts freeze hard.

Campbell decided that once spring arrived, the party would make a march down the coastline of 200 geographical miles (370 km) to Hut Point, but in early September they all were stricken with severe enteritis. By the time they finally left the igloo on 30 September, Browning and Dickason were so weak that they could not help pull the two sledges. Struggling under such conditions, they made less than 30 geographical miles (55 km) in the first week, and any optimism must have been severely challenged. Fortunately, they more than doubled that distance the following week, and they continued slowly to make their way south until, on 29 October, they found a cache of food left by the Second Western Party, which included two containers of biscuits and bags of raisins, tea, cocoa, sugar, butter and lard.

As they moved farther south, they encountered more depots left by earlier parties and they slowly began to regain their strength. Then, on 5 November, they saw other people, and reacted ecstatically. 'Priestley and I walked towards them; they apparently stopped,' Campbell wrote later. 'Priestley started semaphoring while I looked through my glasses. No result. Suddenly they turned and I saw they were Emperor penguins, miraged up in a way that made them look like figures.'

Such disappointment proved insignificant compared to the news they received the very next day, when they reached Hut Point. Their celebration of arriving there was cut short by finding a message from Atkinson, which indicated that his search party had gone south a week earlier to search for the Polar Party. The next day the six men completed their amazing journey to Cape Evans, where they were met by the last two men in camp, the geologist Frank Debenham and the new cook, W.W. Archer. From them, the Northern Party began to learn that they were not the only expedition members who had faced adversity. However, despite their travails, they had triumphed, and lived to tell the tale of one of the most remarkable adventures in the history of polar exploration.

Outside back cover: *Terra Nova* and group photograph SPRI 2005/5/1544 and 2005/5/0368

Inside back cover: Maps showing the outward and inward journeys of the British Antarctic Expedition 1910-13

Published in Great Britain by The Scott Polar Research Institute
University of Cambridge, Lensfield Road, Cambridge, CB2 1ER © 2011

ISBN 978-0-901021-13-7

Design by Dazeye. Printed and bound by Henry Ling Limited, Dorchester.